MATHEMATICS LOVERS
- Math Notebook -

AMANTES DE LAS MATEMÁTICAS
- Cuaderno Cuadriculado -

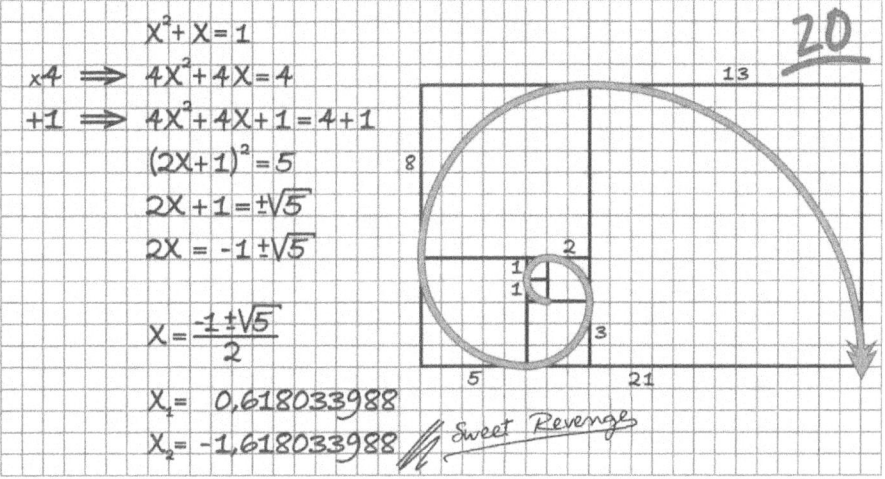

BY
Sweet Revenge®

2017

© **Copyright**

Author: *Sweet Revenge*®
Title: **MATHEMATICS LOVERS** – Math Notebook
Subtitle: AMANTES DE LAS MATEMÁTICAS – Cuaderno Cuadriculado
Year: 2017

PRESENTATION

Mathematics is a universal language that not only humans use it but is also used by chemical elements and molecules that follow mathematical models, many physical phenomena are develop conform to mathematical principles such as the winds and waves of the sea, geomorphology, movements of the stars, planets, satellites and comets follow mathematical patterns. The man did not invent the mathematics because this one is much older than him, but he learned to interpret it and to use it to its own benefit. Nowadays computers are used to perform the most complex calculations. However, until about 50 years ago, mathematicians did their calculations by hand, using pencil and paper.

To learn math is necessary to do calculations by hand and mental. Therefore, this book is a stimulus for you to make calculations by hand, in the style of the old mathematicians like Pythagoras, Archimedes, Newton or Einstein. It does not matter if you only calculate sums, subtractions, multiplication and divisions... or you go further and solve quadratic equations, derivations, integrals... *One of the most important and useful things that everyone should learn is to CALCULATE!*

PRESENTACIÓN

La Matemática es un idioma universal que no solo los humanos lo utilizan, sino que también es utilizada por los elementos y moléculas químicas siguen modelos matemáticos; muchos fenómenos físicos se ajustan a principios matemáticos tales como los vientos y las olas del mar, la geomorfología, los movimientos de los astros, planetas, satélites y cometas siguen patrones matemáticos. El hombre no inventó la matemática porque esta es mucho más antigua que él, pero aprendió a interpretarla y a utilizarla en beneficio propio. Hoy en día se utilizan computadoras para realizar los cálculos más complejos. Sin embargo, hasta hace unos 50 años, los matemáticos hacían sus cálculos a mano, con lápiz y papel.

Para aprender matemáticas es necesario hacer cálculos a mano y mentales. Por eso, este libro es un estímulo para que realices cálculos a mano, a la usanza de los antiguos matemáticos como Pitágoras, Arquímedes, Newton o Einstein. No importa si solo calculas sumas, restas, multiplicaciones y divisiones... o vas más allá y resuelves ecuaciones cuadráticas, derivaciones, integrales... *¡Una de las cosas más importantes y útiles que todo el mundo debe aprender es a CALCULAR!*

Sweet Revenge

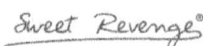

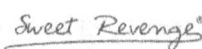

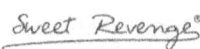

Sweet Revenge

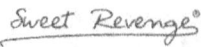

Sweet Revenge

Sweet Revenge

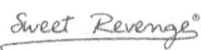

Sweet Revenge

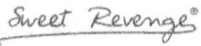

Sweet Revenge

Sweet Revenge

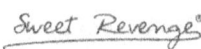

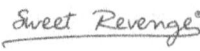

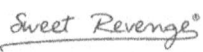

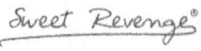

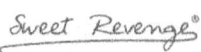

Sweet Revenge

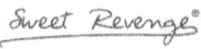

Sweet Revenge

SOME CONSTANTS AND VALUES OF COMMON USE

$\pi \cong 3{,}141593$	$\log \pi \cong 0{,}497149$
$\pi/2 \cong 1{,}570796$	$\ell n 2 \cong 0{,}693147$
$1 \text{ rad} \cong 57{,}29578°$	$\ell n 3 \cong 1{,}098612$
$1° \cong 0{,}017453 \text{ rad}$	$\gamma \cong 0{,}577215$
$e \cong 2{,}718282$	$\sqrt{e} \cong 1{,}648721$
$e^2 \cong 7{,}389056$	$\sqrt{\pi} \cong 1{,}772453$
$1/e \cong 0{,}367879$	$\sqrt{2} \cong 1{,}414213$
$e^\pi \cong 23{,}140692$	$\sqrt{3} \cong 1{,}732050$
$\log 2 \cong 0{,}301029$	$\varnothing \cong 1{,}618033$
$\log e \cong 0{,}434294$	$\Gamma(1/2) \cong \sqrt{\pi}$

SOLUTION OF TRIANGLES

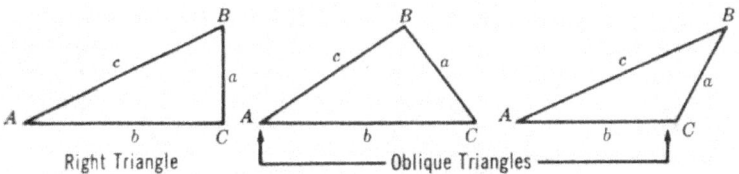

Right Triangle — Oblique Triangles

Solution of Right Triangles

For Angle A. $\sin = \dfrac{a}{c}, \cos = \dfrac{b}{c}, \tan = \dfrac{a}{b}, \cot = \dfrac{b}{a}, \sec = \dfrac{c}{b}, \operatorname{cosec} = \dfrac{c}{a}$

Given	Required	
a, b	A, B, c	$\tan A = \dfrac{a}{b} = \cot B,\ c = \sqrt{a^2+b^2} = a\sqrt{1+\dfrac{b^2}{a^2}}$
a, c	A, B, b	$\sin A = \dfrac{a}{c} = \cos B,\ b = \sqrt{(c+a)(c-a)} = c\sqrt{1-\dfrac{a^2}{c^2}}$
A, a	B, b, c	$B = 90° - A,\ b = a \cot A,\ c = \dfrac{a}{\sin A}$
A, b	B, a, c	$B = 90° - A,\ a = b \tan A,\ c = \dfrac{b}{\cos A}$
A, c	B, a, b	$B = 90° - A,\ a = c \sin A,\ b = c \cos A$

Solution of Oblique Triangles

Given	Required	
A, B, a	b, c, C	$b = \dfrac{a \sin B}{\sin A},\ C = 180° - (A+B),\ c = \dfrac{a \sin C}{\sin A}$
A, a, b	B, c, C	$\sin B = \dfrac{b \sin A}{a},\ C = 180° - (A+B),\ c = \dfrac{a \sin C}{\sin A}$
a, b, C	A, B, c	$A + B = 180° - C,\ \tan \tfrac{1}{2}(A-B) = \dfrac{(a-b)\tan \tfrac{1}{2}(A+B)}{a+b},$ $c = \dfrac{a \sin C}{\sin A}$
a, b, c	A, B, C	$s = \dfrac{a+b+c}{2},\ \sin \tfrac{1}{2} A = \sqrt{\dfrac{(s-b)(s-c)}{bc}},$ $\sin \tfrac{1}{2} B = \sqrt{\dfrac{(s-a)(s-c)}{ac}},\ C = 180° - (A+B)$
a, b, c	Area	$s = \dfrac{a+b+c}{2},\ \text{area} = \sqrt{s(s-a)(s-b)(s-c)}$
A, b, c	Area	$\text{area} = \dfrac{bc \sin A}{2}$
A, B, C, a	Area	$\text{area} = \dfrac{a^2 \sin B \sin C}{2 \sin A}$

GEOMETRIC FORMULAS

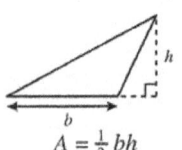
$A = \frac{1}{2} bh$

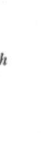
$A = \frac{1}{2} h(b_1 + b_2)$

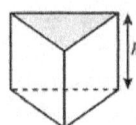
$V = Bh$
$L.A. = hp$
$S.A. = L.A. + 2B$

$V = \frac{1}{3} Bh$
$L.A. = \frac{1}{2} lp$
$S.A. = L.A. + B$

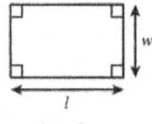
$A = lw$
$p = 2(l + w)$

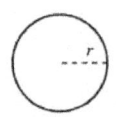
$A = \pi r^2$
$C = 2\pi r$

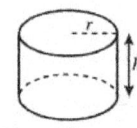
$V = \pi r^2 h$
$L.A. = 2\pi rh$
$S.A. = 2\pi r(h + r)$

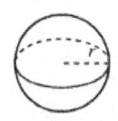
$V = \frac{4}{3} \pi r^3$
$S.A. = 4\pi r^2$

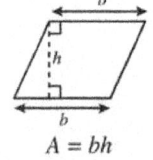
$A = bh$

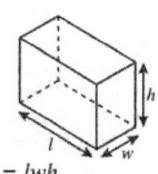
$V = lwh$
$S.A. = 2lw + 2lh + 2wh$

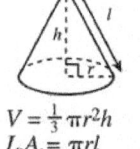
$V = \frac{1}{3} \pi r^2 h$
$L.A. = \pi rl$
$S.A. = \pi r(l + r)$

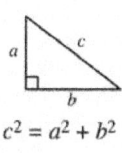
$c^2 = a^2 + b^2$

SOLUTION OF A QUADRATIC EQUATION

$$aX^2 + bX + c = 0$$

$$x = \frac{-b \pm \sqrt{b^2 - 4ac}}{2a}$$

NOTABLE MULTIPLICATIONS

$$(a+b) \cdot (a-b) = a^2 - b^2$$
$$(a \pm b)^2 = a^2 \pm 2ab + b^2$$
$$(a \pm b)^3 = a^3 \pm 3a^2 b + 3ab^2 \pm b^3$$
$$a^3 - b^3 = (a-b) \cdot (a^2 + ab + b^2)$$
$$a^3 + b^3 = (a+b) \cdot (a^2 - ab + b^2)$$

TRIGONOMETRY OF NOTABLE ANGLES

θ	$0°$ / $0°$	$30°$ / $\frac{\pi}{6}$	$45°$ / $\frac{\pi}{4}$	$60°$ / $\frac{\pi}{3}$	$90°$ / $\frac{\pi}{2}$	$180°$ / π	$270°$ / $\frac{3\pi}{2}$	$360°$ / 2π
$\sin \theta$	0	$\frac{1}{2}$	$\frac{1}{\sqrt{2}}$	$\frac{\sqrt{3}}{2}$	1	0	-1	0
$\cos \theta$	1	$\frac{\sqrt{3}}{2}$	$\frac{1}{\sqrt{2}}$	$\frac{1}{2}$	0	-1	0	1
$\tan \theta$	0	$\frac{1}{\sqrt{3}}$	1	$\sqrt{3}$	N.D.	0	N.D.	0
$\csc \theta$	N.D.	2	$\sqrt{2}$	$\frac{2}{\sqrt{3}}$	1	N.D.	-1	N.D.
$\sec \theta$	1	$\frac{2}{\sqrt{3}}$	$\sqrt{2}$	2	N.D.	-1	N.D.	1
$\cot \theta$	N.D.	$\sqrt{3}$	1	$\frac{1}{\sqrt{3}}$	0	N.D.	0	N.D.

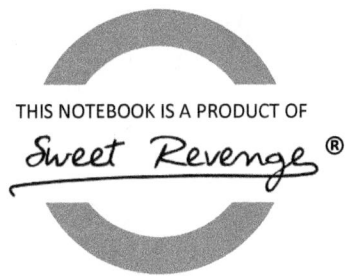

www.ingramcontent.com/pod-product-compliance
Lightning Source LLC
Chambersburg PA
CBHW070239230526
45470CB00002B/456